What's that got to do with me?

Global Warming

Antony Lishak

FRANKLIN WATTS
LONDON•SYDNEY

This edition updated 2009

First published in 2006 by
Franklin Watts
338 Euston Road
London
NW1 3BH

Franklin Watts Australia
Hachette Children's Books
Level 17/207 Kent Street
Sydney NSW 2000

Series editor: Adrian Cole
Design: Thomas Keenes
Art director: Jonathan Hair
Picture researcher: Diana Morris

A CIP catalogue record for this book is
available from the British Library.

ISBN: 978 0 7496 8884 4

Dewey Classification: 363.738'74

Printed in China

Acknowledgements:
Bryan & Cherry Alexander/Alamy: 23, 31. Fred Bavendam/Still
Pictures: 8. Martin Bond/Still Pictures: 14. Cape Wind courtesy of
CapeCodTODAY.com: 12br. Bob Daemmrich/Image Works/Topham: 3-
4,15. Digital Vision: 13. Mark Edwards/Still Pictures: 21. Warren
Faidley/OSF: 7. Raimund Franken/Still Pictures: 9. Eric Gay/AP/Empics:
6l, 16, 28c, 32. A.Ishokon/ UNEP/Still Pictures: front cover b, back cover
b, 6c, 20. ITAR-TASS: 18, 28l. Mark Lynas/Still Pictures: 24, 25, 30. Phil
Noble/PA/Empics: 12l, 29. Novosti/Topham: 19. Sipa Press/Rex
Features: 1, 6r, 22, 26, 27, 28r. J. Tack/Still Pictures: 11.
Woodmansterne/Topham: front cover t, back cover t.

Franklin Watts is a division of
Hachette Children's Books,
an Hachette Livre UK company.
www.hachettelivre.co.uk

Global Warming

Contents

So what?

Gases exist naturally in the Earth's atmosphere. They help trap the heat of the Sun that is so vital for all life on the planet. But over the last century the level of these gases has increased and the Earth is becoming hotter. Scientists have named this process "global warming".

What's it all about?

The effects of global warming, including changes to weather patterns, are caused by gases in the atmosphere, such as carbon dioxide. These are called "greenhouse gases" because they act like the glass that heats air inside a greenhouse. The levels of greenhouse gases have risen as a result of the increased use of oil, coal and gas. These fossil fuels release carbon dioxide into the atmosphere when they are burned.

Many scientists are concerned about the effects of global warming and argue that people must change their energy use to preserve our planet. But some scientists believe global warming is part of the Earth's natural heating process, and not a result of human activity.

Personal accounts

All of the testimonies in this book are true. Some are first-hand accounts, while others are the result of bringing similar experiences together to create a single "voice". Every effort has been made to ensure they are authentic. Wherever possible, permission to use the information has been obtained.

Ask yourself

The testimonies won't tell you all there is to know about global warming; that wouldn't be possible. Instead, as you encounter the different views, think about your own opinions and experiences. This will help you begin to address the question: "Global warming – what's that got to do with me?"

Many scientists believe that global warming will lead to an increase in storms.

A hotel owner

Many believe that, as a result of global warming, ocean temperatures are rising. If this continues it will have a devastating effect on sea life. Howard is an Australian hotel owner whose livelihood depends upon a unique marine habitat.

To the rest of the world the Great Barrier Reef in Australia is a beautiful natural wonder. But Townsville, where I live, and my hotel in particular, depends on the thousands of visitors it attracts each year. If the scientists are right, by 2050 the rising sea temperatures will have killed off the coral that lives on the reef. My kids had better pack up and leave now.

A diver on the Great Barrier Reef.

Fact bank

■ The Great Barrier Reef, off the north-east corner of Australia, is the largest coral reef in the world.

■ High water temperatures kill the algae that live inside coral animals and provide them with much of their food. When the algae die the corals die too. This process is called "bleaching".

■ In a survey during the summer of 2002 scientists found the effects of bleaching on the majority of the Great Barrier Reef.

This aerial view of the Great Barrier Reef shows only part of this huge marine habitat.

The warming waters are killing the coral. Even if global warming was stopped today, and that's not going to happen, it would still take decades for the deadly process to stop. As the habitat and food chains of over a thousand sea creatures disappear, some of the most beautiful life on Earth could become extinct. Forget about trying to find Nemo – he and all the other clownfish won't be there!

Ask yourself this...

■ Why should we care that a non-human environment is being damaged?

■ Can you think of other animal habitats that are being destroyed by global warming?

■ What could be done to stop the destruction of the reef?

A "green" cyclist

Global warming can seem like too much of a problem for one person to make a difference. Simon is someone who believes that if we all change our habits then there is hope.

Most people know there is a problem but they don't believe they can do anything about it. They blame it on "governments" or "industry". But it's our actions that contribute to global warming so we can easily do something to reduce its effects.

People have got to stop wasting electricity, because most of it is still produced using fuels that release greenhouse gases. One simple way to save energy is to stop leaving televisions, DVD players and computers on "stand-by". People could also reduce carbon dioxide emissions if they reduced the number of unnecessary car journeys they make and tried walking or cycling instead. I cycle past hundreds of stationary fuel-guzzling vehicles on my way to work!

You could try taking a shower instead of a bath. Showers use a lot less water – so less gas or electricity is needed to heat up the water.

Simon believes people's habits must change.

Fact bank

■ Cycling is a totally pollution free mode of transport, and it's healthy too!

■ Cars and trucks generate almost 30% of the USA's carbon dioxide emissions.

■ Televisions left on "stand-by" account for 23% of all TV-related energy use.

■ In the USA $3 billion is wasted each year by equipment on "stand-by".

Ask yourself this...

■ Carry out a survey to find out how many children in your class come to school by car. What proportion of them walk or cycle to school each morning?

■ How easy would it be to follow Simon's energy-saving suggestions?

■ What other ways can you think of to reduce energy use?

Reducing energy use helps to cut down emissions produced by power stations.

A rural activist

Wind is a possible carbon-free alternative source of energy that could help to reduce carbon dioxide emissions. But Joyce doesn't think it's such a good idea. She believes that rural areas should not be spoilt by groups of wind turbines – called wind farms.

This sculpture is part of an anti-wind farm campaign.

What's the point of ruining rural areas for the sake of a few megawatts of so called "clean" power? If the future of green energy production involves more of these clumps of wind turbines, then I'd rather have global warming! These ugly eye-sores are the worst architecture in the world.

And it's going to get worse. The global plan to reduce greenhouse gas

These protesters in the USA are campaigning against a new wind farm.

One of many wind farms worldwide.

Fact bank

■ Renewable sources of energy, like wind power, are considered to be the "clean" alternatives to fossil fuels.

■ "Clean" sources of energy do not produce any emissions or other waste products. They include tidal and solar power.

■ Some wind farms are built in the sea and are not visible from land.

■ A recent report concluded that 12% of the world's electricity could be supplied by the wind by 2020.

emissions by 5% by 2012 will mean building more of these awful wind turbines. Where are they going to be built? And who do you think is going to pay for them? Taxpayers like me, that's who!

If you don't live near a wind farm then you just don't understand – they aren't as quiet as people would like to believe. Try lying in bed listening to 60 of them when the wind speed is fairly strong. There must be a better way to save the planet.

Ask yourself this...

■ How deeply do you think Joyce has thought about global warming?

■ What would you say if a wind farm was built near your home?

■ What would you say if a nuclear power station was built near your home? (See pages 26–27.)

■ Why do we need "clean" energy?

A car lover

Carbon dioxide emissions from cars are a major contributor to global warming. But for most drivers it's much easier to say we should drive less than it is to do it. Gordon is a car enthusiast from Detroit, USA.

No one's going to tell me what car I can drive. I'm proud to live in the land of the free. When I get into my SUV [Sports Utility Vehicle] I'm asserting my right as an American citizen to go anywhere I want at any time I please. And being so high up lets me look down on other road users and, let me tell you, when they catch a sight of me in their rear view mirror they sure do make way.

The number one priority is my family's safety and that's what my vehicle guarantees. And what's the alternative? Public transport? Get serious! If you live in the suburbs you just can't survive without a car.

I've heard folk talk about the environment and how the fumes from my car are heating up the planet, but I just don't see the problem. Many scientists doubt there is such a thing as global warming. Who are we supposed to believe?

Car use is increasing throughout the world, not decreasing.

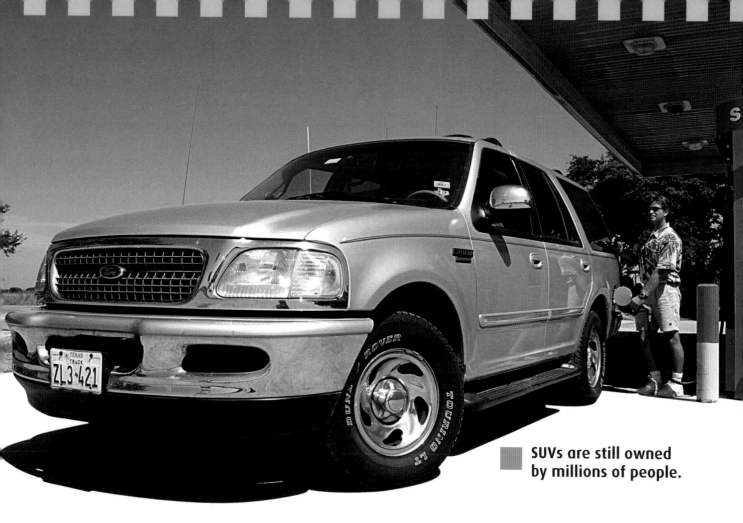

SUVs are still owned by millions of people.

Fact bank

■ SUVs have large engines which are not fuel efficient.

■ A recent survey found that in one year the average amount of energy used by an SUV is the same as leaving a light bulb burning for 30 years or leaving a television on for 28 years.

■ SUVs still account for over 50% of the vehicles owned by people in the USA.

Ask yourself this...

■ This man's priority is his family's safety. So why does he drive a car with high carbon dioxide emissions?

■ Should SUVs be banned in order to force people to drive more fuel-efficient cars? What implications could a ban have for some people?

■ How could public transport be improved? Think about the services in your area. What would encourage people to use them more?

A flood survivor

Many scientists believe rapid climate change has started to alter weather patterns across the globe. Letisha is someone who suffered from the devastating effects of a flood.

Hurricane Katrina turned my world upside down. No one realised how bad the flooding would be in New Orleans. My mum and dad took us to the Superdome when the water started to come into the house. We'd only packed what we could carry because we thought we'd be back soon. There were crowds of people sitting around with nowhere to go.

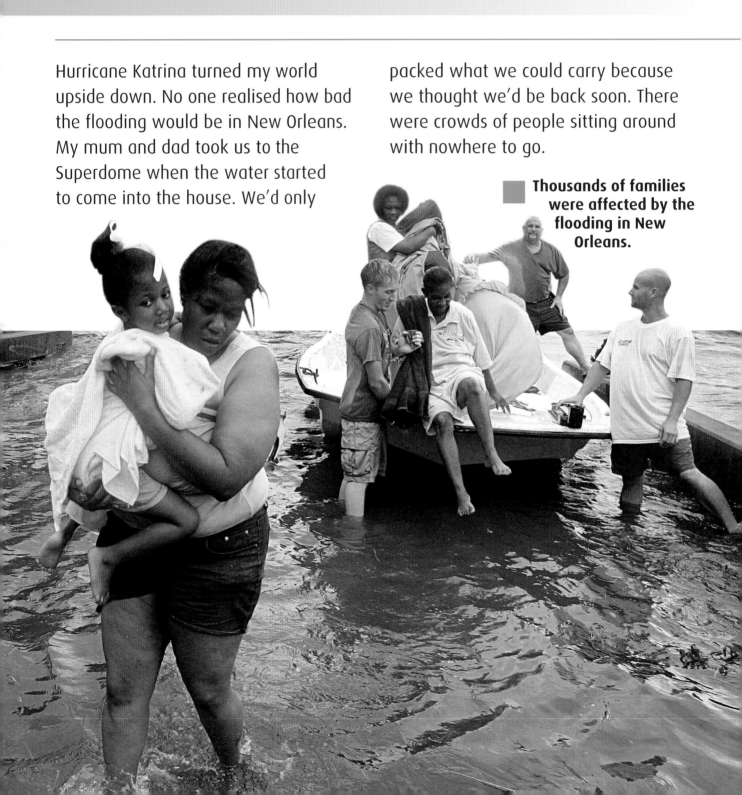

Thousands of families were affected by the flooding in New Orleans.

My dad is a local politician. He had been campaigning since the flood in 1993 for improvements to be made to the levees that control the river. He knew that the whole area was at risk because the city is built below water level. Hurricane Katrina caused waves to surge up the Mississippi River and break through a levee that was designed to keep flood water out of the city. But these levees weren't built for today's bad weather.

I think people had started to become too relaxed about the storms that hit the Gulf Coast. But where else could we go? It was our neighbourhood – but now everything has changed.

Fact bank

■ In the 20th century, average global temperatures rose by 0.6° Celsius. They are expected to rise by up to 5.8° Celsius by 2100, increasing the risk of extreme weather conditions. New Orleans is in a hurricane area and has always been at risk from flooding. Many scientists believe that the effects of global warming will increase these risks.

■ Despite the construction of the Mississippi River control system in the 1870s, the river caused severe flooding in 1927, 1993 and 2005.

Ask yourself this...

■ If the Gulf Coast region is flooded so easily, why do people live there?

■ What do you think could have been done to reduce the risk of flooding in New Orleans?

■ Do you think it is right to link this flood with the effects of global warming? Was it just a natural disaster?

Flood water from the Mississippi River caused widespread devastation.

Timber logging

Trees absorb carbon dioxide and release oxygen, in a natural process called photosynthesis. The health of our planet depends upon the great forests to filter the air for us. But what happens if they are destroyed?

A logger

There is a big demand in my country for wood, and we have to get it from somewhere. What is the problem? Russia has a very big supply of timber. Almost 60% of my country is forest. Some of those trees have been there for 500 years. We could cut

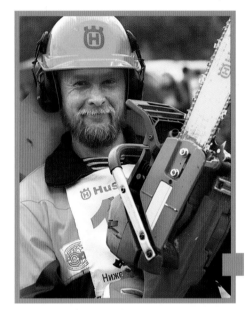

down trees every day for the next 10 years and it would be like snapping a tiny twig from one of these giant pines. It won't make a difference. And in time these trees will grow back again.

A Russian logger ready with his chainsaw.

Fact bank

■ Russia is home to the largest forested region on Earth, with more than 55% of the world's conifer trees. Russia's forests absorb about 10% of global carbon dioxide emissions every year.

■ Forest fires give off large amounts of carbon dioxide. Many are caused deliberately by people clearing away trees for new farmland.

■ Some forests are managed to ensure that for every tree that is chopped down a new one is planted.

■ Every three years an area of Russian forest the size of California disappears.

■ Some forests are being planted to act as "carbon sinks". They are intended to absorb the carbon dioxide emissions produced by some industries.

A local woman

When I was young the rivers were clean. Now the oil from their machines flows in the water. But it's not just the logging – even more of the forest is lost to the fires that logging activities cause. All they care about is money. Of course people need wood for furniture and paper, but it has to be controlled or permanent damage will be done; not just to my land but to the whole world.

Russian timber waiting to be transported downriver.

Ask yourself this...

■ How important is it to control logging across the world? What more could be done to prevent the destruction of forests?

■ Wood is a material that can easily be recycled. What does your family do with paper and wood products that it no longer needs?

■ Can you think of other materials that could be used instead of wood?

An African farmer's son

Many areas of the world, such as Africa, suffer from drought. Some scientists think that global warming is making the problem worse. Here is the voice of a child from North Africa, whose family is struggling to cope.

This child is carrying water from a well.

My father is a farmer and he says that for many generations our ancestors have learned to live with drought. But in the last few years things have got worse. Now our land is turning into a desert.

We grow the food we need to live, but if there is no rain then we can't grow anything. It has not rained properly in my area for three years.

Fact bank

■ The United Nations says that one-third of the Earth's surface is at risk of turning into desert. In Spain 31% of the land is in danger.

■ By 2025, over 60% of farmland in Africa could disappear, along with around 30% of Asian and 15% of South American farmland. Scientists disagree about how much of this will be caused by global warming.

My father says we will have to move to the city. The government doesn't want us to do that. They say there is no room. But what else can we do?

My grandfather blames the farmers who grew more food than they needed to. He says they took all the goodness out of the soil. Some farmers burnt down trees to make new areas for crops but that turned to dry dust when the rains failed again. He says we should stay and that things will improve if we care for the land. But that won't make it rain again.

Ask yourself this...

■ Do you think this family should stay, or go and live in the city?

■ How do you think the effects of global warming are making drought conditions worse?

■ Is global warming the only reason that areas like this are turning into deserts? What other causes can you think of?

Many areas of Central Africa have become desert wastelands.

An Inuit hunter

The most extreme effects of climate change occur in the coldest parts of the world. Here is an Inuit who depends upon the ice remaining frozen.

Global warming is melting my world. Year by year the permafrost that our whole region sits on is thawing into the sea. Areas that were once great expanses of ice have turned into mudslides. We depend upon the ice. It has been the source of our survival for centuries. Now it is thinning and

The Arctic ice cap is melting rapidly.

huge cracks make it almost impossible to navigate the hunting grounds that have been used for generations. Experienced hunters have been stranded as ice has floated away from the mainland and drifted out to sea. As the ice breaks so does the food chain that we are all part of. The seal colonies have reduced and there are fewer lemmings and foxes, which means that polar bears are losing their main source of food. Now we humans are considering leaving.

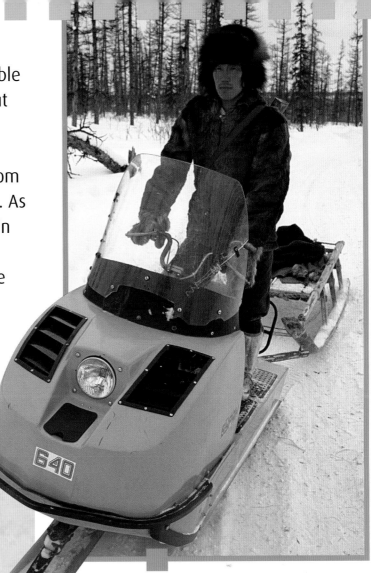

An Inuit hunter on his skidoo.

Fact bank

■ When permafrost melts, huge amounts of carbon dioxide, trapped in the soil below, are released into the atmosphere.

■ Antarctica has lost 13,000 square kilometres of ice-sheet over the past 50 years.

■ Arctic sea ice lost 43% of its thickness between 1976 and 1996.

■ The melting of the Greenland ice sheet would raise sea levels by approximately 6 metres.

■ Some scientists have warned that polar bear numbers could drop by over 30% by 2050.

Ask yourself this...

■ How much should you care that a few people who live in a remote, cold part of the world might have to move somewhere else?

■ Who, if anyone, should take responsibility for the melting of the ice caps and the rising of the seas?

■ How might rising sea levels affect the rest of the world?

An island inhabitant

Melting ice causes the oceans to rise. This spells disaster for those people who live on land that is only just above sea level, as this island inhabitant is discovering.

Fact bank

■ Tuvalu is situated in the Pacific Ocean about 3,200 kilometres north-east of Australia.

■ Tuvalu's highest point is only 4 metres above sea level.

■ New Zealand has agreed to resettle 75 people a year from Tuvalu.

The sea has already damaged land on Tuvalu.

My island, Tuvalu, is likely to be the first state in the world to be submerged by rising sea levels. Scientists say it'll be uninhabitable by 2025, and by the end of the century it will have disappeared altogether.

Already, stretches of beach are being swallowed up by the waves and the saltwater of the ocean is rotting the roots of trees. Many crops have been destroyed and we are dependent on imported, canned food. Last year, without warning, the sea bubbled up

Waves crash against coastal homes on Tuvalu.

through cracks in the ground, tearing up the roads and flooding our homes.

My whole world is dying around me. It makes me feel sad and angry at the same time. Sad that eventually we will have to leave our home, and angry because this has been caused by the actions of others who only look after their own needs and don't see the bigger picture.

Ask yourself this...

■ If you were an inhabitant of Tuvalu what would you do?

■ What could be done to help people whose land is almost at sea level?

A nuclear supporter

Jean-Paul works in a nuclear power station in France. He believes that the answer to reducing greenhouse gases is quite straightforward. The solution to the world's increasing demand for energy lies in the production of nuclear power.

Some day we will have burned all the world's fossil fuels. Of course, it won't happen tomorrow, but it is going to happen. Who knows what damage fossil fuel emissions will have done to the planet by then? And our need for electricity is going to keep growing.

It seems pretty simple to me. We have to embrace "clean" nuclear power.

I know a lot of people have reservations about its use, but

French politicians celebrate the future of electricity production by nuclear power.

Fact bank

■ In 1979 mechanical and human error in the nuclear power plant at Three Mile Island, Pennsylvania, caused dangerous radioactive gases to escape. The building of nuclear power plants in the USA stopped.

■ A recent report shows that nuclear power is the second-cheapest way to generate electricity, at 2.3p per kilowatt hour (kw/h), after gas at 2.2p. Wind power costs more than 5p per kw/h.

A nuclear power station in France.

■ France is the world's second-biggest producer of nuclear power. Over 75% of its electricity is produced in this way. It is estimated that this has cut France's carbon dioxide emissions over the last 30 years by 50%

■ A by-product of nuclear power is radioactive waste, which must be reprocessed or stored for hundreds of years. The world's largest reprocessing plant is in La Hague, France.

technology is so much safer today. There are organisations, such as the United Nations International Atomic Energy Agency (IAEA), watching over safety, too.

By the year 2050 there will be approximately 10 billion people living on this planet. All of them are going to need power. Nuclear power is the only reliable, immediately available energy source that could meet their needs in full. And the waste is controllable. There are no carbon dioxide emissions from a nuclear power station.

Ask yourself this...

■ How accurate is it for this person to call nuclear power "clean"?

■ How can nuclear power be seen as "better" than power generated by burning fossil fuels?

■ Given that accidents have happened in the past, how confident are you that "safe" nuclear power stations can be built?

What does global warming have to do with me?

Greenhouse gases in the atmosphere are at their highest level for at least 420,000 years. The temperature of our planet has risen 1° Celsius in the last 500 years – half of that has happened in the last century. Many scientists believe that global warming will affect everyone on Earth. The debate about what to do is less clear. Should we use "clean" energy sources or nuclear power? Should we change the way we lead our lives to reduce carbon dioxide emissions? Or is it too late to do anything about it?

Global warming quiz

Here are some questions you can answer to help you think about what global warming has to do with you. Compare your score with that of your friend's, then go back through the book. Use all this information to form your own opinion about global warming.

1. Do you always turn lights off when you leave an empty room?

2. Does your home have energy-saving light bulbs fitted?

3. Do you turn your television off completely when it is not being used, and not leave it on "stand-by"?

4. Do you take showers not baths?

5. Does your family recycle?

6. Do you walk or cycle to school, only using the car when there is no real alternative?

7. Does your family own a fuel-efficient car?

8. Does your family buy products that are advertised as "energy efficient"?

9. Have you ever planted a tree?

10. Do you use public transport when you can?

If you answered "yes" to less than 3 out of 10:

Read this book again and look at the websites on pages 30–31. Talk to your friends and family to discuss what global warming has to do with all of you.

If you scored between 4 and 8 out of 10:

You've clearly thought about what global warming has to do with you, but there's still more you can do.

If you scored more than 8 out of 10:

You really know how your actions affect the environment. Talk to other people who didn't score so highly about the other things they can do.

Websites

The websites below feature more information, news articles and stories that you can use to help form your own opinions. Use the information carefully and consider the source it comes from before drawing any conclusions.

www.greenpeace.org
Homepage of Greenpeace International, with links to sections on climate change, nuclear energy and forest protection, plus news, games and campaign stories.

www.windpower.org/ en/core.htm
Website of the Danish Wind Industry Association, including news and information about offshore wind farms in Denmark.

www.worldwildlife.org/ climate
Climate-change section of the Wildlife Wildlife Fund website with an introduction to global warming, and lots of conservation information, including a "What can you do" section.

www.reefed.edu.au
Australian government reef education website featuring images and the Great Barrier Reef Explorer.

www.unep.org/themes/ climatechange
Website of the United Nations Environment Programme featuring lots of climate-change information from around the world.

www.nrdc.org/ globalwarming
Natural Resources Defense Council website with information about all aspects of global warming.

Glossary

www.iaea.org
Website of the International Atomic Energy Agency, featuring news and stories from the nuclear industry.

**http://epa.gov/
   climatechange/kids/
   index.html**
Kids' website of the US Environment Protection Agency website, with climate information, history, games and movies.

www.climatechange.gov.au
"Greenhouse" Office of Australia's Department of the Environment and Heritage, including energy-saving tips and information about global warming.

Every effort has been made by the Publishers to ensure that these websites contain no inappropriate or offensive material. However, because of the nature of the Internet, it is impossible to guarantee that the contents of these sites will not be altered. We strongly advise that Internet access is supervised by a responsible adult.

Carbon dioxide – one of the gases found naturally in the Earth's atmosphere. The build up of carbon dioxide is believed to be causing a rise in air temperatures, which many scientists call global warming. Carbon dioxide is released when fossil fuels are burned to generate energy.

Drought – a long period without rain.

Emission – the action of something being released. When fossil fuels are burned they emit carbon dioxide.

Enthusiast – someone with a strong liking for something.

Food chain – a series of animals and plants in a "chain" that feed on each other and depend on those sources of food for survival. For example, seaweed ▶fish ▶seals ▶polar bears.

Fossil fuels – these are natural materials made from the remains (fossils) of long-dead creatures that are burned to generate energy. They include natural gas, coal and oil.

Fuel efficient – using fuel with the minimum amount of waste or emissions.

Global warming – the increase in average temperatures across the world, thought to be caused by the greenhouse effect.

Green – a casual term used to describe someone who campaigns for the environment.

Greenhouse effect – the way the Earth's atmosphere traps heat inside it, like the glass of a greenhouse.

Greenhouse gas – a gas in the Earth's atmosphere, such as carbon dioxide, that traps heat from the Sun.

Permafrost – land that is usually permanently frozen over, often to great depths.

Public transport – the bus and rail networks run to transport people from one place to another.

Rural – in the countryside, not in the city.

Index